Scott Wilken

Big Buddy Books

An Imprint of Abdo Publishing
abdobooks.com

abdobooks.com

Published by Abdo Publishing, a division of ABDO, PO Box 398166, Minneapolis, Minnesota 55439.

Printed in the United States of America, North Mankato, Minnesota
052025
092025

Design: Elena Klinkner, Mighty Media, Inc.
Production: Mighty Media, Inc.
Editor: Ruthie Van Oosbree
Cover Photograph: Meindert van der Haven/iStockphoto
Interior Photographs: Anton Starikov/Adobe Stock, p. 28 (baking pan); apomares/iStockphoto, p. 7 (cirrus clouds); BillionPhotos.com/Adobe Stock, p. 28 (ice); Carsten Reisinger/Adobe Stock, p. 28 (ruler); Dany Schmalz/iStockphoto, p. 17 (roll cloud); Dashu Xinganling/Shutterstock, p. 5; Edward Bilcliffe/iStockphoto, pp. 24–25; esen/Adobe Stock, p. 28 (flashlight); Gerville/iStockphoto, p. 7 (nimbus clouds); Jason Benz Bennee/Shutterstock, pp. 18–19; Jason Persoff Stormdoctor/Getty Images, p. 27; JrGarcia/iStockphoto, p. 11; kornienko alexandr/Adobe Stock, p. 28 (spray can); MarBom/iStockphoto, p. 7 (cumulus clouds); Meindert van der Haven/iStockphoto, pp. 12–13; Mick Petroff/Wikimedia Commons, p. 21; Mighty Media, Inc., p. 29; nikol85/Adobe Stock, p. 28 (black paper); olaser/iStockphoto, p. 17 (shelf cloud); Orbs production/Shutterstock, pp. 14–15; Thiago Orsi Laranjeiras/Shutterstock, pp. 8–9; Uwe Michael Neumann/Shutterstock, p. 23; Wayward_hiker1/Shutterstock, p. 7 (stratus clouds); Winai Tepsuttinun/Adobe Stock, p. 28 (jar)
Design Elements: Mighty Media, Inc.

Library of Congress Control Number: 2024948549

Publisher's Cataloging-in-Publication Data
Names: Wilken, Scott, author.
Title: Curious clouds / by Scott Wilken
Description: Minneapolis, Minnesota : Abdo Publishing, 2026 | Series: Weather wonders | Includes online resources and index.
Identifiers: ISBN 9781098296360 (lib. bdg.) | ISBN 9798384917793 (ebook)
Subjects: LCSH: Clouds--Juvenile literature. | Atmosphere--Juvenile literature. | Sky--Juvenile literature. | Weather--Juvenile literature.
Classification: DDC 551.6--dc23

Contents

Many Kinds of Clouds

All clouds are made up of ice crystals or water drops. But they can look very different from one another. Some clouds are thin and **wispy**. Others are fat and puffy. Clouds are usually white or gray. But they can be other colors too.

The science of studying clouds is called nephology.

Scientists have found four main types of clouds. These are **cirrus**, **cumulus**, **stratus**, and **nimbus**. Most clouds are one of these types. But some clouds don't fit into a category. Others are unusual **versions** of one or more types.

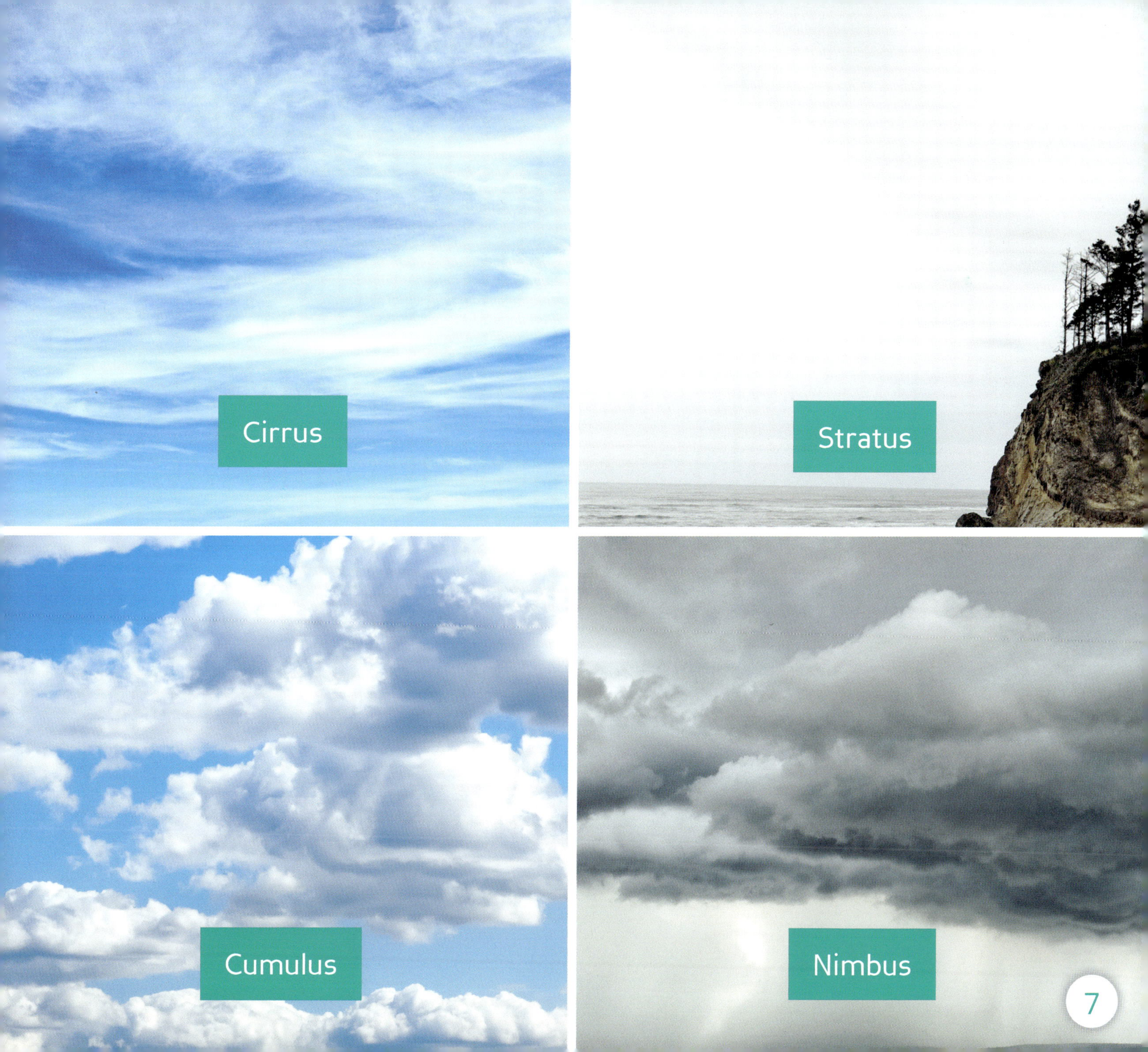
Cirrus
Stratus
Cumulus
Nimbus

Lenticular Clouds

Lenticular clouds are an uncommon type of cloud. These usually form near mountains. When air blows over a mountain, it can force a cloud into flat, layered shapes. A lenticular cloud may look like a stack of pancakes.

Many reported unidentified flying objects (UFOs) turned out to be lenticular clouds.

Mount Shasta in California is known for often having lenticular clouds. The clouds may make the mountain seem **supernatural**. There is a story that a city of advanced beings called Lemurians exists inside the mountain.

Some people think that the lenticular clouds at Mount Shasta hide the Lemurians' spaceship!

Mammatus Clouds

Mammatus clouds are another unusual type of cloud. They often form underneath **cumulonimbus** clouds. Air sinking within a cloud may cause **bulges** to form below it. These bulges are the mammatus cloud. They often occur right before or after thunderstorms.

Mammatus clouds look especially cool when the sun shines on them.

Arcus Clouds

Arcus clouds usually occur during strong thunderstorms. They form when cold air hits the ground and spreads outward. This cold air pushes warmer air up toward the storm cloud. As the warm air rises, drops of water in the air create an arcus cloud.

When arcus clouds form, there are often strong winds, thunder, and lightning.

There are two types of arcus clouds: shelf clouds and roll clouds. The rising air usually reaches the storm cloud above. This is a shelf cloud. But wind may blow in different directions above and below the rising air. This causes the air to spin, forming a separate roll cloud.

Shelf cloud
Roll cloud

Morning Glory Clouds

Morning glory clouds are very uncommon. They are most often seen over the Gulf of Carpentaria in northern Australia around September and October each year. Morning glory clouds are shaped like long tubes. They sometimes occur in groups.

A morning glory cloud can be nearly 600 miles (966 km) long!

Morning glory clouds form in the early morning. They are created when two sea breezes meet. The air where the breezes meet moves in waves. Moisture in the air on one side of the wave becomes a cloud.

Morning glory clouds form low in the sky.

Nacreous Clouds

Nacreous clouds occur near the north and south poles. They form very high in the sky when the temperature falls below about –120 degrees Fahrenheit (–84.4°C). The cold causes tiny water drops to freeze into **wispy** clouds of ice crystals.

Sunlight hitting the ice crystals makes nacreous clouds look very colorful.

Nacreous clouds can be harmful to Earth. The ice crystals in the clouds cause **chemical reactions** that increase harm to the **ozone layer**. Also, the ice crystals take in heat from Earth. Some of the heat is sent back to Earth, warming the surface. Both of these effects play a part in **climate change**.

Nacreous clouds are sometimes called mother-of-pearl clouds.

Wondrous Weather

Clouds constantly move and change. This makes them fun and interesting to watch. But they can also be scary when they bring strong storms. Have you ever seen any of these uncommon and wonderful clouds?

Wavelike Kelvin-Helmholtz clouds form when the top layer of clouds moves faster than the bottom layer.

Make a Cloud in a Jar!

What You Need

- ruler
- glass jar
- hot water
- ice cubes
- metal pan
- hair spray
- flashlight (optional)
- black paper

What You Do

1. Have an adult pour 2 inches (5 cm) of hot water into the jar.
2. Put ice cubes in the metal pan.
3. Spray hair spray in the jar. Quickly set the pan of ice on top of the jar.

4. Watch a cloud start to form in the jar! Try turning off the lights and shining a flashlight on the jar to see it even better.
5. After a few minutes, hold the black paper behind the jar and remove the pan of ice. The cloud will rise out of the jar!

Glossary

bulge—something that sticks out in a rounded lump.

chemical reaction—a scientific process in which two or more things combine to form something new.

cirrus (SEER-uhs)—a type of cloud that occurs high in the sky and has a wispy appearance.

climate change—long-lasting change in Earth's weather and temperature patterns.

cumulonimbus (KYOO-myoo-loh-NIM-buhs)—a type of cloud that reaches high into the sky and produces thunderstorms.

cumulus—a type of cloud that usually has a flat base, clear outline, and fluffy appearance.

nimbus—a cloud that carries rain, snow, or hail.

ozone layer—a gas layer that surrounds Earth and blocks out certain sun rays.

stratus—a type of cloud that occurs low in the sky and has a broad appearance.

supernatural—not existing in the natural world.

version—a different form or type of an original.

wispy—fine, light, and airy.

Online Resources

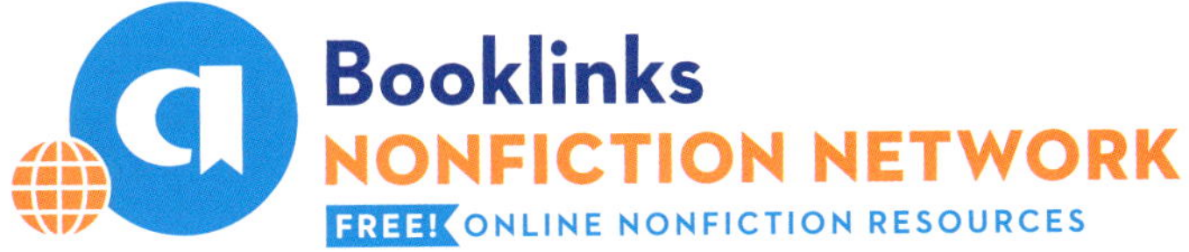

To learn more about clouds, please visit **abdobooklinks.com** or scan this QR code. These links are routinely monitored and updated to provide the most current information available.

Index